SHARK FILES

LOS ARCHIVOS DEL TIBURÓN

GREAT WHITE SHARKS

TIBURONES BLANCOS

Written by / Escrito por: Julie K. Lundgren
Translated by / Traducción de: Sophia Barba-Heredia

TABLE OF CONTENTS

ÍNDICE

A Crabtree Seedlings Book
Un libro de El Semillero de Crabtree

Crabtree Publishing
crabtreebooks.com

School-to-Home Support for Caregivers and Teachers

This book helps children grow by letting them practice reading. Here are a few guiding questions to help the reader with building his or her comprehension skills. Possible answers appear here in red.

Before Reading:

- What do I think this book is about?
  - *I think this book is about great white sharks.*
  - *I think this book will tell me what they eat.*
- What do I want to learn about this topic?
  - *I want to learn where great white sharks live.*
  - *I want to learn how many teeth they have.*

During Reading:

- I wonder why...
  - *I wonder what tools are used to study great white sharks.*
  - *I wonder why people kill sharks for their teeth.*
- What have I learned so far?
  - *I have learned that each shark's fin is unique like fingerprints.*
  - *I have learned that they swim in both deep and shallow waters.*

After Reading:

- What details did I learn about this topic?
  - *I have learned that great white sharks are the deadliest hunters in the sea.*
  - *I have learned that they catch and eat rays, sea turtles, and California sea lions.*
- Read the book again and look for the vocabulary words.
  - *I see the word* ***lurk*** *on page 6 and the word* ***dorsal*** *on page 16. The other glossary words are on pages 22 and 23.*

Apoyos de la escuela a los hogares para cuidadores y maestros

Este libro ayuda a los niños en su desarrollo al permitirles practicar la lectura. Abajo están algunas preguntas guía para ayudar al lector a fortalecer sus habilidades de comprensión. En rojo hay algunas opciones de respuesta.

Antes de leer:

- ¿De qué pienso que tratará este libro?
  - *Pienso que este libro es sobre tiburones blancos.*
  - *Pienso que este libro me dirá qué comen.*
- ¿Qué quiero aprender sobre este tema?
  - *Quiero aprender dónde viven los tiburones blancos.*
  - *Quiero aprender cuántos dientes tienen.*

Durante la lectura:

- Me pregunto por qué...
  - *Me pregunto qué herramientas son usadas para estudiar a los tiburones blancos.*
  - *Me pregunto por qué las personas matan a los tiburones para obtener sus dientes.*
- ¿Qué he aprendido hasta ahora?
  - *Aprendí que cada aleta de tiburón es única como las huellas dactilares.*
  - *Aprendí que ellos nadan tanto en aguas profundas como en aguas no profundas.*

Después de la lectura:

- ¿Qué detalles aprendí de este tema?
  - *Aprendí que el tiburón blanco es el más mortífero cazador del océano.*
  - *Aprendí que ellos atrapan y comen rayas, tortugas de mar y lobos marinos de California.*
- Lee el libro de nuevo y busca las palabras del glosario.
  - *Veo la palabra* ***acechan*** *en la página 6 y la palabra* ***dorsal*** *en la página 16. Las demás palabras del vocabulario están en las páginas 22 y 23.*

GREAT WHITE SHARKS

Which fish is the deadliest hunter in the sea?

TIBURONES BLANCOS

¿Qué peces son los cazadores más mortíferos del océano?

Great white sharks!
They catch big **prey**.

¡Los tiburones blancos! Ellos atrapan **presas** grandes.

ray

raya

sea turtle

tortuga de mar

California sea lion

lobo marino de California

FROM THE FILES

Rays, sea turtles, and California sea lions are prey for great whites.

DESDE LOS ARCHIVOS

Rayas, tortugas de mar y lobos marinos de California son presas de los tiburones blancos.

Great whites **lurk** below, then speed to their prey above.

Los tiburones blancos **acechan** desde abajo, luego avanzan rápidamente hacia su presa, que está arriba.

They can feel movement from **distant** animals.

Pueden sentir el movimiento de animales **distantes**.

FROM THE FILES

Great whites can see and hear well.

DESDE LOS ARCHIVOS

Los tiburones blancos pueden ver y oír bien.

Sharp **teeth** slice food. Gulp!

Con sus filosos **dientes** rebanan la comida. ¡Gulp!

They swim in both deep and **shallow** waters.

Nadan tanto en aguas profundas como en aguas **poco profundas**.

Tools help us study sharks.

Hay herramientas que
nos ayudan a estudiar
a los tiburones.

Each shark has a **dorsal** fin that is not like any other shark's.

Cada tiburón tiene una aleta **dorsal** que no es como la de ningún otro tiburón.

FROM THE FILES

We know each shark by its fin, like a fingerprint!

DESDE LOS ARCHIVOS

Reconocemos a cada tiburón por su aleta, ¡como una huella dactilar!

People kill sharks for their jaws, teeth, and fins.

Hay personas que matan a los tiburones para obtener sus mandíbulas, dientes y aletas.

There is much more to learn about great white sharks.

Hay mucho más por aprender sobre los tiburones blancos.

GLOSSARY / GLOSARIO

distant (DISS-tehnt): Distant animals or objects are far away.

distantes: Los objetos o animales distantes son los que están lejos.

dorsal (DOR-suhl): A dorsal fin is the fin on top of a shark's back.

dorsal: Una aleta dorsal es la aleta que está en lo alto de la espalda del tiburón.

lurk (LERK): To lurk is to hide in sneaky silence.

acechan: Que se esconden en cauteloso silencio.

prey (PRAY): Prey is any animal hunted and eaten by another animal.

presas: Las presas son animales cazados y comidos por otros animales.

shallow (SHAL-oh): Shallow water is not deep and is often near land.

poco profundas: Las aguas poco profundas no son hondas y usualmente están cerca de tierra firme.

teeth (TEETH): Teeth are white, bony parts inside a mouth that are used for biting and chewing.

dientes: Los dientes son las partes blancas y huesudas dentro de la boca, son usados para morder y masticar.

Index

Índice analítico

Websites
Sitios web (páginas en inglés):

www.montereybayaquarium.org/animals/animals-a-to-z/white-shark
www.natgeokids.com/uk/discover/animals/sea-life/great-white-sharks

About the Author

Julie K. Lundgren grew up near Lake Superior where she reveled in mucking about in the woods, picking berries, and expanding her rock collection. Her interests led her to a degree in biology. She lives in Minnesota with her family.

Acerca de la autora

Julie K. Lundgren creció cerca del Lago Superior, donde se divertía jugando en el bosque, recogiendo moras y expandiendo su colección de rocas. Sus intereses la llevaron a hacer una carrera en Biología. Vive en Minnesota con su familia.

Crabtree Publishing

crabtreebooks.com 800-387-7650

Hardcover 978-1-0398-6926-4
Paperback 978-1-0398-6905-9
Ebook (pdf) 978-1-0398-6947-9
Epub 978-1-0398-6968-4

Printed in the U.S.A./Impreso en los Estados Unidos/CP112025

Library and Archives Canada Cataloguing in Publication
Available at the Library and Archives Canada

Library of Congress Cataloging-in-Publication Data Available at the Library of Congress

Published in Canada
Publicado en Canadá
Crabtree Publishing
616 Welland Avenue
St. Catharines, Ontario
L2M 5V6

Published in the United States
Publicado en los Estados Unidos
Crabtree Publishing
347 Fifth Avenue
Suite 1402-145
New York, NY 10016

Written by/Escrito por: Julie K. Lundgren Designer/
Diseñadora: Jennifer Dydyk
Editor/Editora: Kelli Hicks
Proofreader/Correctora de pruebas: Melissa Boyce
Translated by/Traducción de: Sophia Barba-Heredia
Spanish-language layout and proofread/
Maquetación y corrección en español: Base Tres
Production manager/Coordinadora de producción:
Candice Campbell

Photographs: Photographs: Shark illustration on cover logo © BATKA/Shutterstock; white shark illustration for "FROM THE FILES" © Dashikka/Shutterstock; Cover photo © Sergey Uryadnikov/Shutterstock; Page 3 © RamonCarretero/istock; page 4 (ray and sea turtle) © richcarey, (sea lion) © GlobalP/istock; page 5 © USO/istock; page 7 © Sergey Uryadnikov/Shutterstock; page 9 © Ramon Carretero/Shutterstock; page 11 © Martin Prochazkacz/Shutterstock; page 12 © Willyam Bradberry/Shutterstock; page 13 © Brent Barnes | Dreamstime.com; page 15 (top) © Stefan Pircher/Shutterstock, (bottom) © OCEARCH/R. Snow; page 17 © Sergey Uryadnikov/Shutterstock; page 18 © Alessandro De Maddalena/Shutterstock; page 19 © Andrey Simonenko | Dreamstime.com; page 21 © USO/istock; page 22 (top photo) © Aerial-motion/Shutterstock